Geo F. Bailey

Descriptive and pictorial illustrations of the wild animals

mentioned in natural history

Geo F. Bailey

Descriptive and pictorial illustrations of the wild animals mentioned in natural history

ISBN/EAN: 9783741194603

Manufactured in Europe, USA, Canada, Australia, Japa

Cover: Foto ©Klaus-Uwe Gerhardt /pixelio.de

Manufactured and distributed by brebook publishing software (www.brebook.com)

Geo F. Bailey

Descriptive and pictorial illustrations of the wild animals mentioned in natural history

DESCRIPTIVE AND PICTORIAL

ILLUSTRATIONS

OF THE

WILD ANIMALS,

MENTIONED IN

NATURAL HISTORY.

PRICE 25 CENTS.

NEW YORK:

PRINTED BY S. BOOTH, 199 & 201 CENTRE STREET.

1870.

To the Reader.

HE study of Natural History has of late years not received that attention, from the general public, that its great importance demands; and, indeed, the apathy is not very surprising, for the obvious reason that there are but few works upon this interesting study put into such form that they can *cheaply* reach the youth of the country. To supply the want, the proprietors of this menagerie have thought it advisable to publish an abridgment of the subject, — in which a succinct description of the various subjects of the Animal kingdom is given, and many of which are on exhibition in this institution. There is, certainly, nothing more pleasing, either to adults or the young, than a comprehensive history of the nature and habits of the wild quadrupeds of the forest and the jungle.

The limited space of a little work of this kind, of course, precludes the possibility of entering otherwise than briefly on the theme; but it is hoped that it may prove to be an incentive to pursue the study and the means of obtaining a scientific knowledge of Natural History. So important has this branch of study become in Europe, that the authorities of many of the principal cities and seats of learning—particularly in England, France, and Germany—have deemed the founding of Zoological Institutes an imperative necessity, in order that the students in the various colleges and normal schools might have opportunities to compare ocular observation with the statements of the several writers on Zoology. These institutions are subsidized by their respective governments, and have agencies established in Asia, Africa, South America, and other parts of the globe, for the purpose of securing and transporting any rare or extraordinary beast or reptile that may come under observation.

Our descriptions have been compiled from standard works on the subject, and it is hoped that they may have a tendency to lead the reader into a contemplation of the vastness and sublimity of the animal world, and show the wisdom and potency of the Great Creator of the universe, and all that therein is.

Many of the animals on exhibition in this Menagerie are the finest of their species,—in fact, they have been pronounced by those well versed in Natural History, to be the best specimens ever seen on this Continent.

Description of Wild Animals.

THE ELEPHANT.

THIS animal is the most gigantic of existing quadrupeds. It is characterized essentially by having grinders composed of alternating plates of ivory, enamel and cæmentum, and two tusks in the upper jaw,—it is also the only living Mammalia which has a proboscis, or trunk, longer than the head. There are two species of elephants—the Indian and African. The former differs in its conformation from the latter in its greater size, in the skull being higher in proportion to its length, and with a more concave forehead. The Indian has also comparatively smaller ears, the skin is of a paler brown color, and it has four nails on the hind feet, instead of three. The height of the Indian elephant, measured from the top of the shoulder has rarely been found to exceed ten feet six inches,—the ordinary height being from seven to nine feet. The tusks of the Indian elephant varies as to length; their ends only are visible externally in the female. The characteristics of the African elephant may be inferred from those of the Indian species. It is usually described as having a forehead convex instead of concave, but the projection is caused by the nasal bones, which are higher placed than in the Indian species, and the true front is, in reality, concave in the African, but in a less degree than in the Indian. The chief external character of the African elephant is his huge ears, which descend to his legs.

The elephant has small eyes, compared with his prodigious size, but they are sensible and lively, and manifest a pathetic, sentimental expression. He is reflective and deliberative, and never acts or moves until he has carefully observed the signs that he is to obey. When tamed, he is the most tractable and submissive of all quadrupeds, and evinces great affection for his keeper. He has a tenacious memory, always remembers a kindness, but never forgets or forgives an injury or a practical joke. He has a quick ear, and a very acute sense of smell, which is more marked in him than in any other animal. He delights in music, learns to keep good time, and moves in cadence to it. His sense of feeling is particularly sensitive, and centres in his trunk, which is composed of membranes, nerves and muscles ;— he can not only move or bend it, but can shorten, lengthen, and turn it in ever conceivable shape. The extremity of this trunk terminates by an edge, which projects above, like a finger, and with which he can pick from the ground the smallest piece of money ; he can gather flowers, and exhibits a fine discrimination in choosing them ; unties knots and unlocks doors—in short, this sagacious animal can be taught to do almost anything.

The ordinary walk of the elephant is not quicker than that of a horse; but when pushed, he assumes a kind of ambling pace, which in fleetness is equal to a gallop. He goes forward with great case and celerity, but it is with great difficulty that he turns himself round, and that not without taking a pretty large circuit. He swims well, and is of much use in carrying baggage over large rivers. When swimming, he raises his trunk above the surface of the water for the purpose of respiration, every other part of his body being below. In this manner several of these animals swim together, and steer their course without danger of running foul of each other.

It is a singular circumstance in the history of this wonderful animal, that in a state of subjection, it is unalterably barren; and, though it has been reduced under the dominion of man for ages, it has never been known to breed, as if it had a proper sense of its degraded condition, and obstinately refused to increase the pride and power of its conquerors by propagating a race of slaves.

THE LION.

THIS is the largest, most formidable and noble of the carnivorous animals, though not the most typical of the genus at which it stands at the head. It is chiefly distinguished by the presence of a full, flowing mane in the male, by a tufted tail and a disappearance of the feline markings, in both sexes, before reaching maturity; the color then being a nearly uniform light fulvous brown, with mane inclining to black, especially in the Central and South African races. The mane is scantier and lighter colored in the Asiatic than in the African lions. Compared with other animals of his species, the lion combines more robustness with the feline attributes; and his pre-eminent stature receives an air of nobility from the mane that decorates his head and neck. He has a credit, too, of having a greater share of boldness and generosity than the other cats. His vocal organs also exhibit a modification of structure not present in other felines, by which he has the power to utter his tremendous roar, —a roar which, when sent forth under the excitement of hunger, scares from their hiding places the timid ruminants which may be lurking within the compass of its fearful reverberations. The lion, which traverses the parched deserts of Africa, and lies in wait to intercept the Antelopes which bound in troops from one oasis to another, would be rendered too conspicuous if his tawny hide were ornamented by the stripes or spots that characterize the feline livery; these, therefore, which are obvious enough in the earlier periods of his existence, become obliterated as he attains to maturity.

The form of the lion is strikingly bold and majestic; his large and shaggy mane, which he can elevate at will, surrounding his awful front; his huge eye-brows and large and fiery eye-balls, which seem to glow with peculiar lustre, together with the formidable appearance of his teeth, exhibit a picture at once terrific and grand, and which no words can adequately describe.

The length of the largest lion is between eight and nine feet, the tail about four, and its height about four feet and a half. The female is smaller and without the mane. The lion seldom attacks any animal openly, except when compelled by hunger— in that case no danger will deter him; but as most animals endeavor to avoid him, he is obliged to have recourse to artifice, and take his prey by surprise. For this purpose he crouches on his belly in a thicket, where he waits till his prey approaches; and then, with a spring, he leaps upon it at a distance of fifteen or twenty feet, and generally seizes at the first bound. However, if he misses his object, he relinquishes the pursuit; and turning back towards the place of his ambush, he measures the ground, step by step, and again abides his time for another opportunity. The lurking place of the lion is generally near a spring of water, or by the side of a stream, where he frequently has an opportunity of catching such animals as may come to slake their thirst. The lion is a long lived animal, although naturalists have widely differed as to the precise period of its existence. Buffon limits it to twenty-two years; but this is evidently erroneous, as it has been known to live beyond that age. The great lion named "Pompey," which died in the Tower of London in 1760, was known to be in that historic "institution" over seventy years, and another one died in the same place some years since at the age of sixty-three.

The attachment of the lioness to her young is manifested to a remarkable degree. She usually conceals her cubs in retired and inaccessible places, and, when afraid of her retreat being discovered, endeavors to hide her track by brushing the ground with her tail.

ROYAL BENGAL TIGER.

T HE Tiger is one of the most destructive and rapacious of the carnivorous animals. Its thirst for blood is insatiable, and is the most cruel and fierce of its species; however much glutted with slaughter, he will not desist in the work of destruction so long as a single object remains,—flocks and herds fall indiscriminate victims to his rapacity; he fears not the opposition of man, whom he too often makes his prey—in fact, he prefers human flesh to that of any other animal. As his name indicates, this tiger is a native of Asia, but is met with in greater numbers in the East Indies than in any other part of this great division of the globe. In fact, they are so numerous in the remote parts, that often whole districts have to be abandoned, so much is his approach dreaded by the natives, who call him the man stealer; and, for the purpose of his extermination, immense hunts are organized, composed of elephants, and large bodies of horse and foot soldiers. To effect the capture of this dreaded beast, large nets are stretched out, the jungle is surrounded, the tall grass set on fire, and the tiger is driven with shouts and other noises towards the nets, where he is shot at from little houses, which are constructed for the purpose on trees, or on strong poles, or from the backs of elephants. The Bengal Tiger is one of the many wild animals that cannot be subdued to the will of man—in captivity, his temper is not softened either by constraint or kindness—he seems insensible to the attention of his keeper, and is as ready to tear the hand that feeds as he is that by which he is chastized. His strength is prodigious, and can carry off a calf or a deer with such apparent ease that it seems to be no impediment to the rapidity of his flight. He seldom pursues his prey, but springs upon it from a place of ambush, with elasticity, and from a distance of fifteen or twenty feet. He will attack all kinds of animals, even the lion—between whom frequent combats have beem maintained, and in which both have been known to perish. The usual length of the tiger is from five to six feet, exclusive of the tail, which is about two and a half feet long. His symmetry is perfect, and his appearance is majestic. He has a rounder head than the lion—is clothed with short hair, and has no mane. His coat is of a tawny color on top and sides, and white below, with irregular crossed black stripes over the body, neck and sides of the head. A traveler gives an account of a battle between a tiger and two elephants in Siam, of which he was an eye witness. He says: "The arena was enclosed by a high palisade of bamboo canes. One of the elephants approached the tiger, which was confined by cords, and dealt him two or three heavy blows with its trunk, and the tiger lay for some time as if he were dead; but, notwithstanding this attack had a good deal abated his fury, he was no sooner relieved from the cords than with a fearful roar he made a spring at the elephant's trunk, which that animal dexterously avoided by drawing it up, caught the infuriated tiger upon its tusks and threw him into the air. Both elephants were then allowed to come up, and would have undoubtedly killed the royal beast if an end had not been put to the combat." Under such restraints, it is not surprising that the tiger was worsted in the combat; but it can be imagined of what power and fierceness this animal is endowed with, when, after being disabled by the first attack of the elephant, while bound with cords, he would venture to continue such an unequal engagement.

THE LEOPARD.

THIS animal is found over Africa, in Arabia, and the East Indies. Its length, from nose to tail, is about four feet; the body is of a light yellow, and the spots with which it is diversified are smaller and closer than those of the panther, to which animal it bears a great resemblance. The leopard abounds in the interior parts of Africa, from whence they come down in large numbers and make sad havoc among the herds that cover the plains of Lower Guinea, and when beasts of chase fail they spare no living creature, so sanguinary are they in nature. The negroes capture them in pit-falls, slightly covered at the top, and baited with meat. The flesh of the leopard is said to be white, and as well tasted as veal, which is the inducement for the natives hunting them. In India there is a species of leopard, about as large as a greyhound, which is called the *Chetah,* or hunting leopard. It has a small head and short ears; its face, chin and throat are of a pale brown color, inclining to yellow; the body is of a light tawny brown, covered with small round black spots over the back, sides, head and legs; the hair on the top of the neck is longer than the rest; the belly is white; the tail is very long, and marked on the upper side with black spots. For the purpose of hunting, it is carried in a small kind of wagon, chained and hoodwinked until it is allowed to pursue the game. It begins the chase by creeping along with its belly to the ground, stopping and concealing itself till it gets a favorable position, when it darts upon its prey with amazing agility.

THE JAGUAR, or BRAZILIAN TIGER.

HIS animal belongs to South America, and differs greatly from the Asiatic tiger in the color and markings of his skin The hair on his head, back, and outward side of the legs is of a reddish-yellow—not streaked with black, like the royal tiger, but varied with large black annular marks, each having one or more dots in the centre; the belly and inner side of the legs are whitish, with dark transverse stripes. The Jaguar is the king of the predatory quadrupeds in the hot countries of S. America, and is hunted in various ways. Dogs, generally, are set on his track, who follow him into the thicket, whither, glutted with prey, he has retired to rest. As soon as he perceives the dog, he endeavors to get into a tree that is inaccessible to the dog, and from this place of fancied security he is shot down by the hunter. The jaguar is often caught by means of pit-falls, running knots, and spring-guns. In Paraguay it is captured with the lasso, which a horseman, galloping past, throws round its neck; he then drags the animal about till it is exhausted. Sometimes the hunters attack it armed only with lances seven feet long. They plant the weapon firmly in the ground, holding it with both hands; the jaguar, irritated by dogs, springs upon the hunter who must catch it on the spear; but if he fails his life will surely be sacrificed, unless he has time and presence of mind to plunge a long knife into a vital part of the infuriated animal.

The jaguar is a solitary animal, residing in forests, especially near large rivers. It is an expert climber and an excellent swimmer; it has been known, after having destroyed a horse, to swim with the carcass over a deep and broad river without any apparent signs of fatigue. Although ferocious in its wild state, the jaguar in captivity becomes tame and tractable, and is fond of licking the hands of those with whom he is familiar.

THE COUGAR.

THE Cougar, or, as it is often called, the Puma, is the largest of the feline species found in North America. It has also been termed the American lion, though it bears not the least resemblance to the "jungle king." The cougar measures from four to five feet long, exclusive of the tail, which is half the length of its body, and is about two feet and a half high. Its color is of a reddish brown, intermixed with black; the ears and extremity of the tail are dark brown; the under jaw and throat are of a reddish white, and on the abdomen the hair is long and ash-colored. The agility of this animal in climbing and leaping is astonishing, as, when it pursuit of prey, it often springs a distance of twenty feet, from one tree to another. It preys on sheep, calves, young horses, or any animal it can master, and has been known, when near human habitations, to carry off children. The cougar has also been known to tear open as many as fifty sheep at one time, not for the purpose of devouring their flesh, but in order to suck the blood, of which he is very fond, and which has such an intoxicating influence over him, that as soon as the appetite is satiated, he falls asleep on the spot. The method of seizing his prey is by crawling stealthily on his belly through shrubs and bushes until he can reach his victim, when he suddenly springs on its back and tears it to pieces. He is cowardly in his mode of attack, and generally springs from behind his victim.

THE OCELOT.

THIS animal is found in Brazil and other parts of S. America. It is voracious, but timid, and seldom attacks men; if pursued by dogs it seeks safety in flight to the woods. It resembles the domestic cat in conformation, altho' much larger—being nearly three feet long and about two feet high. Its haunts are chiefly in the mountains, where it conceals itself in the thick foliage of the trees, and, when an opportunity presents itself, darts upon such birds or small quadrupeds as come within its reach. The ocelot is strategic, too, for it often stretches itself along the limb or branch of a tree, as if dead, for the purpose of inveigling monkies, who, instigated by their inordinate curiosity, are sure to fall into the snare. The skin of the male Ocelot is remarkably beautiful and most elegantly variegated. Generally, its color is a bright tawny—a black stripe extends along the back from head to tail; its forehead is spotted with black, as also are its legs; the shoulders, sides and rump are beautifully marbled with long stripes of black, forming oval figures, filled in the middle with small black spots. The colors of the female are not so bright as those of the male, nor is she so beautifully marked. The ocelot retains his natural ferocity in captivity, and is rarely tamed. It would appear that it is not susceptible of docility, and no kindness can soften the ferocity of its disposition or calm its restlessness.

THE PANTHER.

THE Panther is next in size to the tiger, and, so little difference is there in appearance, naturalists have often mistaken one for the other. It is a native of Africa, but is also found in the hot parts of Asia, the islands of the East Indies, in Persia, and Northern China. Its hair is short and smooth, and, instead of being striped like the tiger, is beautifully marked on the back, sides and flanks with black spots, disposed in circles of from four to five in each, with a single spot in the centre, thus affording some resemblance to the form of a rose,—on the face, breast and legs they are single; the color of its body is of a deep yellow on the back, graduating to a lighter shade towards the belly, which is white; has short and pointed ears and a restless eye. It is an untamable animal, and fierce and cruel in its nature. The manner of taking its prey is by surprise—lurking in thickets, or creeping on its belly till it comes within springing distance.

When goaded by the pangs of hunger, it will attack, without distinction, any living creature, but prefers the flesh of quadrupeds to that of mankind; it will even climb trees in pursuit of monkeys and lesser animals, so that nothing is secure from its rapacity. The manner in which the panther is captured by the Caffres in Africa is ingenious. They hang a piece of flesh upon a tree, at a moderate height from the earth, and in the bushes underneath is fastened an upright stake, with a sharp point: the panther springs at the bait, and as it comes to the ground is impaled on the concealed weapon. The panther is about the height of a large mastiff, but its legs are not quite so long.

THE LYNX.

THE Lynx, though of the cat kind, differs very much from the species already described. It has long been famed for its sharp sight, and is distinguished by its short tail and tufted ears, which are long and erect. The hair on its body is long and soft, of a red-ash color, and marked with dusky spots, which differ according to the age of the animal. Its legs and feet are thick and strong—its eyes are of a pale yellow color. The skin of the male is more spotted than that of the female. The lynx is a very destructive animal, but less ferocious than the panther; is reputed long-lived, and lives by hunting; pursues its prey to the tops of the highest trees, and delights in the flesh of squirrels, weasels, etc., which rarely escape its pursuit. It lies in wait, concealed in a tree, for the approach of fallow deer, hare, and the smaller quadrupeds generally; it seizes its victim by the throat, and after drinking the blood abandons the carcass and goes in search of fresh game. Its sight is remarkably keen and quick, and espies its prey at a long distance. It is asserted that it will eat no more of a goat or a sheep than the brain, liver and intestines. The red lynx is found in all northern parts of the world, and is hunted for its skin, which is greatly esteemed for its soft thick fur, and which is largely exported from Northern Europe and America. When attacked by a dog, it lies on its back, strikes desperately with its claws, and frequently vanquishes its assailant. Its manner of howling is not very dissimilar to that of the wolf, for whom it is often taken, but there is nothing in common with the two animals.

THE STRIPED HYENA.

THE Striped Hyena is a native of Barbary, Egypt, Abyssinia, Nubia, Syria, Persia and the East Indies. So striking are the characteristics of this animal, that it is hardly possible to be deceived by them. It is, perhaps, the only quadruped which has only four toes to either the fore or hind feet; its ears are long, straight, and nearly bare; its head is more square and shorter than that of the wolf; its legs, especially the hind ones, are longer; its eyes are placed like those of the dog; the hair and mane are of a brownish grey, with transverse dark brown or blackish bands on the body. Its height ranges from eighteen to twenty five inches, and its usual length, from the muzzle to the tail, is a little over three feet. It generally resides in the caverns of mountains, in the clefts of rocks, or in dens which it has formed for itself under the earth. Like the wolf, it lives by depredation, but is stronger and more daring than that animal. It will attack man, carries off cattle, breaks into sheepcots at night, and ravages with an insatiable ferocity. When at a loss for prey, and stimulated by the pangs of hunger, it will scrape up the earth with its feet, and devours the remains of both animals and men, which, in the country it inhabits, are interred promiscuously in the open fields.

THE SPOTTED HYENA.

SOUTHERN Africa is the native country of this animal, and it is found in large numbers in the districts contiguous to the Cape of Good Hope, where it is called the tiger-wolf. It has the same propensities as the Striped Hyena, but is smaller in size. The color of its skin is a dirty yellow, the whole body is covered with blackish brown spots, excepting the under part of the belly and breast, the inner side of the limbs, and the head. The muzzle is black, and the tail is covered with long bushy hair, of a blackish brown. The jaws, like those of the striped genus, are of enormous strength, with which he breaks to pieces the hardest bones. It is generally supposed that this animal is untamable, but instances are on record of both species having been

domesticated, and manifesting as much attachment to man as
has the dog.   In captivity, the Hyena exhibits an irritable and
restless disposition, but that is unquestionably attributable to its
great aversion to a pent up confinement.

THE CAMEL.

THIS quadruped is especially organized for existence in the
arid and barren deserts of Asia and Africa.   The complete
adaptation of the Camel for the dreary wastes in which it
is destined to exist is fully illustrated in its peculiar structure.
It has a broad expanded foot, elastic as a cushion, terminated in
front by two comparatively small hoofs, and well defended beneath
by a felt of coarse hair, which prevents the leg from sinking in
the loose surface; while its long joints and lofty tread adapt it
for a rapid progress along the loose sandy plains.   But the distinguishing characteristic of the camel is its faculty of abstaining
from water for a greater length of time than any other animal,—
for which Nature has made a wonderful provision in giving it,
besides the four stomachs which it has in common with other
ruminating animals, a fifth bag, serving as a reservoir for water,
where it remains without corrupting or mixing with other ali-
ments.   When pressed with thirst, and has occasion for water to
macerate its food while ruminating, it makes part of it pass into
the stomach by a simple contraction of certain muscles.   Besides

this reservoir of water to meet the exigencies of long journeys across the deserts, where streams and vegetation are scarce, the camel is provided with a storehouse of solid nutriment, on which it can draw for supplies long after every digestible part has been extracted from the stomach: this storehouse consists of one or two large collections of fat, stored up in ligamentous cells, supported by the spines of the dorsal vertebræ, forming what is called the hump. When the camel is in a fertile region, the hump becomes plump and expanded; but after a protracted journey in the wilderness it becomes shrivelled and reduced, in consequence of the absorption of the fat. Thus to the Arab of the burning seas of sand the camel is as valuable, and indeed as essential, as the reindeer to the Laplander in his region of perpetual snow. The one animal, like the other, serves for all the purposes of draught and burden. When dead, the flesh of the camel is eaten—though coarser than that of the ordinary ruminants. Its hide, which approaches that of the pachyderms in thickness and strength, is applied to the manufacture of saddles, harness, shields, and various other articles. The finer hair is manufactured into articles of clothing, and the coarser hair is woven into a kind of matting for the covering of tents. By day, the camel carries its master and his family, with their property, from place to place; while at night the body of the recumbent beast of burden serves as a pillow for its owner.

The camel is the only medium of communication between those countries which are separated by extensive deserts. In the expressive and beautiful metaphor of oriental speech, it is the "Ship of the Desert,"—in truth, it is the only transport by which the dreary wilderness of sand can be navigated with safety and certainty.

THE GIRAFFE.

THE GIRAFFE is the tallest quadruped on the face of the earth, and the largest and most singular of the ruminant order. It is a native of the unfrequented deserts of Ethiopia and other interior parts of Africa, where it leads a solitary life, far from the habitations of men. The height of this extraordinary animal, from the crown of its head to the ground, is

about seventeen feet, while at the rump it measures but nine—
the neck alone exceeds six feet. Its head, which resembles that
of the camel, is symmetrical, and undoubtedly the most attrac-
tive part of its structure. The large, dark and lustrous eyes of
the giraffe, which beam with a peculiarly mild but fearless expression, are so placed as to take in a wider range of the horizon
than is subject to the vision of any other quadruped; and while
browsing on its favorite acacia tree, by means of its laterally projecting orbits, can direct its sight so as to anticipate any threatened attack in the rear from the stealthy lion or any other foe of
the desert. To an open attack it sometimes makes a successful
defence, by striking out its powerful and well armed feet, which
are cloven; and it is recorded that the lion has been frequently
repelled and disabled by the wounds which the giraffe has thus
inflicted with its hoofs. It has small horns, muffled by skin and
hair, which are its chief means for defence, and are by no means
the insignificant weapons that their appearance would indicate.
But, notwithstanding those natural arms of hoofs and horns, the
giraffe will not turn to do battle, except as a last extremity:
when escape is possible it seeks it in flight. Its pace on rising
ground is extremely rapid, but its endurance is not such as to
maintain a long chase with a well mounted hunter.

The general figure of the giraffe, its raised anterior parts,
elongated neck, light and tapering head, and long, slender and
flexible tongue, are all conditions which beautifully harmonize
with its geographical position, and the nature of its food. No
ruminant of its magnitude could exist in the arid tropical region
to which this animal is peculiar, if it were not modified so as to
be able to obtain vegetable sustenance independently of ordinary
pasturage. But in those localities, shrubs and trees continue to
put forth buds and leaves when all the herbage on the surface of
the earth is scorched; and it is for the purpose of browsing on
the green food supplied by lofty branches that the ruminant type
is modified in so extraordinary a manner, as is witnessed in the
conformation of the giraffe. It is difficult to acclimate this animal, and, when in captivity out of its native clime, rarely lives
beyond a period of two or three years.

THE ZEBRA.

HIS beautiful animal inhabits the Southern parts of Africa, where large herds of them are often seen grazing on the extensive plains that lie towards the Cape of good Hope. The Zebra is the most elegantly clothed of all quadrupeds. It has the shape and graces of the horse, and the swiftness of the stag. It is larger than the ass, and rather resembles the mule in shape; its head is large, and the ears are long; its legs are small and well placed; its body is rotund, fleshy and well formed; its head is striped with fine bands of black and white, which form a centre in the forehead; its neck is adorned with stripes of the same color running round it, and the body is beautifully variegated with bands running across the back and ending in points at the belly; its thighs, legs, ears, and even the tail, are all beautifully streaked in the same manner. The zebra is about seven feet long, from the point of the muzzle to the origin of the tail, and about four feet high. Such is the beauty of this creature, that it seems by nature fitted to gratify the pride, and formed for the service of man. Its liberty has remained uncontrolled, and its natural fierceness has as yet resisted every attempt to subdue it. In fact, many of them which are in captivity have exhibited so much viciousness that rendered it unsafe to approach them with too much familiarity. There is an animal called the Quagga, which has been confounded with the zebra, but it is a distinct species, although bearing a strong resemblance to it.

THE STAG.

THE Stag is the most beautiful animal of the deer species, and is of so mild, innocent and tranquil a disposition that it seems as if it were created solely to adorn and animate the solitude of the primeval forests, and to occupy, remote from human habitations, the peaceful retreats of nature. The elegance of its form, its flexible yet nervous limbs, its bold branching horns, which are annually renewed; while its size, swiftness, and strength places it pre-eminently before any other beast of the forest. The age of the stag is known by its horns. The first year exhibits only a short protuberance, which is covered with a hairy skin; the next year the horns are straight and single; the third year produces two antlers, the fourth year three, the fifth year four; and when arrived at the sixth year the antlers amount to six or seven on each side, but the number is not always certain. The stag begins to shed its horns at the latter end of February or beginning of March. Stags in the seventh year do not undergo the change till the middle or end of March; nor do those in their sixth year till the month of April. Soon after the old horn has fallen off, a soft tumor begins to appear, which is soon covered with a down-like velvet; this tumor buds forth every day like the graft of a tree, and, rising by degrees, shoots out the antlers on each side. The most common color of the stag is yellow, though there are many found of a brown and many of a red color. The Hind conceals its young with great care, in obscure retreats, and will expose herself to the fury of the hounds in order to draw them off from the place where she has concealed her Fawn.

THE REINDEER.

THE Reindeer inhabits Norway, Lapland, and the northern regions of Asia and America. It bears a resemblance to the stag, which it equals in size; but it has shorter and stouter legs, and the hair is thicker round the neck; its horns are from three to four feet in length, curving backwards from the base and returning towards the top, as illustrated in the engraving. The color of its skin, in summer, is a greyish brown, the lower part of the neck and belly being white; in winter, however, it becomes lighter, and the hair is longer. In the summer season the reindeer feeds on grass, leaves and buds, and during the winter months it exists upon a peculiar *lichen*, called reindeer moss, which it digs up with its horns and hoofs, often from under several feet of snow. The flesh of the reindeer is excellent eating, and its tongue, when smoked, is held in high esteem. This animal is employed by the inhabitants of the northern regions in drawing their sledges, and, so rapid is their speed, they frequently accomplish journeys of one hundred miles in a day. The wild herds of reindeer, when migrating, always form in long files and keep close together, the females invariably taking the lead. This extraordinary animal stands in the same relation to the inhabitants of the frigid climates as does the camel to those of the arid regions of the orient,—for it supplies the place of oxen, sheep, and horses, and furnishes good milk, butter and cheese; while its hide is used for clothing, tents, beds, etc., and mattresses and cushions are stuffed with the hair; their bones are worked up into spoons, knives, and various other articles.

THE ELK.

OF the deer species, this is the largest and most formidable, and is found in Europe and America. It is known in the former country as the Elk, and in in the latter as the Moose. It also inhabits Norway, Sweden, Poland, and Russia, and is common in Canada and all the northern parts of this continent. The horns of this animal present a singular appearance, as they are not divided into branches, but grow together in a fan-like shape, and terminate in a number of points. In summer, its color is of a dark brown; inclining to a greyish white on the lower parts, and in winter a grey brown; its hair is rough and shaggy, and on its back there is a short upright mane. Its head is long and rudely formed, and its upper lip projects three inches over the lower one; its nostrils are very broad, however, its sense of smell is not acute, but its sight and hearing are much better. Although the elk is an inoffensive animal, it will, when attacked, defend itself with great vigor by means of its horns and fore feet, with which it can deal very severe blows. The flesh of the elk is palatable, and much relished by the Indians and white settlers in the fur domains in the north-west; it bears a greater resemblance to ox beef in its flavor than to venison

THE NYL GHAU.

THIS interesting quadruped is a native of India. It appears to be of a middle nature between the cow and the deer, and bears the appearance of both in its form. In size, the Nyl Ghau is much smaller than the one and larger than the other; its body, tail and horns are not unlike those of the bull, and the head, neck and legs are similar to those of the deer. The general color of the animal is ash or grey, from a mixture of black hairs and white; along the ridge of the neck and back the hair is blacker, longer, and more erect, making a short, thin, and upright mane, reaching down to the hump; its horns are seven inches long, six inches round at the root, tapering by degrees, and terminating in a blunt point; the ears are large and beautiful, seven inches in length, and spread to a considerable breadth, —they are white on the edge and inside, except where two black bands mark the hollow of the ear with a zebra-like variety. The height of this animal at the shoulder is about four feet one inch : behind the loins it measures only four feet. The female differs from the male, both in height and thickness, she being much smaller, and in shape and color resembling the deer, and has no horns.

THE AXIS DEER.

THE Axis, or Indian Stag, is a native of Asia, and is found in the immense plains that are watered by the Ganges. Its distinguishing features from both the stag and fallow deer are the large white spots, elegantly disposed and distinct from each other, over its body, which is of a light brown color ; its horns are round, like those of the stag, but it has no antlers. The axis appears to be an intermediate mixture between the stag and the deer. It resembles the deer in the size of its body, the

length of its tail, and the color of the skin—the only essential
difference being in the horns, which, as before stated, resemble
those of the stag.   The axis, therefore, may possibly be only a
variety depending upon the climate, and not a different species of
deer—for, though it belongs to the hottest countries of Asia, it
supports and easily multiplies in those of Europe.

THE HART-BEEST.

THIS animal is the most common of all the larger gazelles
known in Africa. The height of the Hart-Beest to the
top of the shoulder is about four feet, and its form is between that of the stag and the heifer. Its color is of a dark cin-
namon, the front of the head and fore part of the legs being
marked with black; the hind part of the haunch is covered with
a wide black streak, which reaches to the knee. Its height exceeds four feet. Its horns are from six to nine inches long, very
strong and black, almost close at the base, diverging upwards,
and at the top bending backwards in a horizontal direction almost
to the tips, which turn downwards, and are embossed with rings
of an irregular form. The hair of the hart-beest is very fine,
and its ears are covered with white hair on the inside; it has
only eight teeth in the lower jaw and none in the upper; its legs
are slender, with small fetlocks and hoofs. This animal has a

large head and high forehead, which, with its asinine ears and tail, render it one of the least handsome of, the whole of the antelope species. Its pace, when at full speed, appears like a heavy gallop, but, nevertheless, it runs as fast as any of the larger antelopes. When pursued, and well ahead of pursuers, it is apt to turn round and gaze at them. Like the nyl-ghau and wood antelope, it drops on its knees to fight. As an article of food it is agreeable in flavor, though rather dry.

THE ZEBU.

THIS quadruped is a haunched ox, and is the ordinary domestic animal throughout India, Persia, Arabia, and in Africa from the Atlas to the Cape of Good Hope. The Zebu varies in size and color, and some of the species are without horns. Its general color is of a pale grey, or white; and between the shoulders there is a fleshy hump, which varies in weight, ranging from forty to fifty pounds. There are numerous varieties of this animal, ranging in size from a large mastiff dog to that of a full grown buffalo. The zebu, in the countries above stated, supplies the place of the ox, both as an article of food and a beast of burden. In parts of India it also performs the duties of a horse —being either saddled and ridden, or yoked to a vehicle; they make long journeys with considerable celerity, and it is stated by travellers that they can accomplish from twenty to thirty miles in a day. The zebu is held in high veneration by the Hindoos, who consider it a sacred animal, and, consequently, look upon its slaughter, and the eating of its flesh, as an act of impiety and desecration; but they do not scruple to make the animal labor for their benefit, although they generally exempt a select number from service, and permit them to stray through villages and towns, where they procure food from pious contributors or others who impose upon themselves the charitable function.

THE IBEX.

THIS animal resembles the domestic goat in appearance, and is a native of the Alps, Pyrenees, and other mountains in Europe. The head of the Ibex is small, adorned with large curved horns, covered with nobs, and which, in some instances, have been found to exceed two yards in length. Its color is of a deep brown, mixed with ash; a streak of black runs along the top of its back, and the belly and thighs are of a delicate fawn color. The female is a third smaller than the male, with smaller horns, which rarely exceed a length of eight inches. The male ibex differs from the chamois by the length, thickness and form of the horns: it is also more bulky, vigors and strong; in other respects the two animals have the same manners and customs. The ibex is covered with a firm and solid skin, which, in winter, is covered with a double fur, with rough hair outwardly and a finer and thicker hair underneath. The ibex, when taken young and brought up with domestic goats, is easily tamed, imbibes the same peculiarities and herd together.

THE GRIZZLY BEAR.

THE Grizzly Bear inhabits, principally, the northern part of America, but is found in greater numbers in the neighborhood of the Rocky Mountains, where, on its 'native heath,' it reigns as supreme as does the lion in the sandy wastes of Africa. Its muscular strength is so great that few animals of the forest or the plains care to have any dealings with him at close quarters—not excepting the bison, who is a foe worthy of his hug. In size, the grizzly bear is about double that of the black bear, although, in general appearance, they resemble each other. Its feet are enormously large—the breadth of the fore foot exceeding nine inches, and the length of the hind one, exclusive of the talons, being eleven inches, and its breadth seven. The color of its hair varies from a light grey to a dark brown—the latter shade being predominant. However, it is always, in some degree, qualified by an intermixture of greyish hairs, the brown hairs only being tipped with grey. Its hair is longer, finer, and more exuberant than that of the black bear. The grizzly bear is remarkably tenacious of life: an instance is on record of one of these animals, being followed by a party of hunters, having received as many as fourteen balls into different parts of its body, and lived an hour or two after. This animal can exist on vegetable food alone, but it has a great fondness for destruction and blood.

THE POLAR BEAR.

THIS animal differs very materially from the common bear in the length of its head and neck, and attains about twice the size—often measuring thirteen feet. Its limbs are of immense size and strength; its hair is long, harsh, and disagreeable to the touch, and of a yellowish-white color; its ears are short and rounded, and has large teeth. The White or Polar Bear has rarely been seen further south than Newfoundland, but it abounds chiefly on the shores of Hudson Bay, Greenland, and Spitzbergen. It has also been found in Norway and Iceland, whither it is very often carried on floes of ice. During the summer it takes up its residence on an island of ice, and frequently passes from one to another. It swims well, and often goes a distance of six or seven leagues; it also dives, but cannot remain

long under water. At sea, the food of the polar bear is fish, seals
and carcasses of whales; on land, it preys upon deer and other
animals, and will eat various kinds of berries. In winter, it beds
itself deeply under the snow, or eminences of ice, and awaits, in
a torpid state, the return of the sun. It suffers greatly when ex-
posed to intense heat. Of the ferocity of the polar bear there are
many thrilling instances related. A few years since a party of
sailors, in a boat, fired at and wounded one, when it immediately
swam after the boat, and, overtaking it, endeavored to climb into
it, and while one of his feet was on the gunwale it was cut off
with a hatchet; but, notwithstanding the severe wounds, he still
pursued the sailors to the ship. Numerous additional wounds
did not check its fury or prevent its ascending to the deck of the
ship, where it was finally killed.

THE LARGE-LIPPED BEAR.

THE body of this animal is covered with a long, rough and
shaggy coat of hair, which, when lying down, gives it the
appearance of a rude and shapeless mass; on the top of its
back the hair, which is twelve inches long, rises up like a hunch,
separates in the middle, and falls down in different directions;
its head is large and broad at the forehead, being the only part
on which the hair is short, and the snout is of a yellowish-white.
The tail is so short as to be scarcely visible. Its lips are thin,

very long, and furnished with muscles, by which it can protrude them in a most singular manner; its eyes are small, black and heavy, and its aspect lowering; its legs and thighs are remarkably thick and strong, which resemble those of the common bear; it has five long, crooked white claws on each foot, which it uses with great dexterity, either separately or together, like fingers, to break its food and convey it to the mouth. It is a gentle, good natured animal, but, when irritated or disturbed, utters an abrupt roar, ending in a whining tone, expressive of impatience. This curious animal is a native of the interior parts of Bengal; it burrows in the ground, and feeds on nuts, fruit, honey, or bread.

THE BROWN BEAR.

THIS is a savage and solitary animal, and is found in almost every climate. It inhabits the most unfrequented parts, and chooses its den in the most gloomy parts of the forest, in some cavern that has been hollowed by the workings of time, or in the hollow of an enormous tree. It subsists chiefly on roots, fruits and vegetables, but is occasionally carnivorous. It retires to its den in the latter part of autumn (when it is exceedingly fat) and lives throughout the winter in a state of inactivity, without provisions, or without stirring abroad in quest of food. It seems to sustain itself upon the plenitude of fat that it has accumulated during the summer months, and only feels the cravings of hunger when its adipose has considerably wasted. When this occurs, which it generally does at the expiration of forty or fifty days, the male forsakes its den; but the female remains confined for four months, during which time she brings forth her young. Her retreat for that purpose is chosen in the most retired places, apart from the male, lest he should devour the cubs, which he certainly would do if an opportunity presented itself. The bear is remarkably fond of honey, for which it seeks with cunningness and avidity, and will encounter almost any difficulty to procure it. Its senses of seeing, hearing, and feeling are remarkably acute. It has small eyes, short ears, and a coarse skin.

THE WOLF.

THIS animal, both externally and internally, seems to resemble the dog, but without any of its characteristics; indeed they are so different in their dispositions that no two other animals have such an antipathy to each other. A young dog will shudder at the sight of a wolf; but a dog that is strong, and knows his strength, will attack the wolf with courage and avidity, and endeavors to rid himself of an object that he considers his natural enemy. They seldom meet without fighting or fleeing from each other. Should the wolf prove the stronger, he tears and devours his prey; the dog, on the contrary, is more generous and rests content with its victory. The wolf when captured young becomes tame, but never shows any attachment, and with age its natural dispositions are resumed, and it returns to the woods from whence it was taken. The wolf as it grows old gets greyer, and its teeth wear by using. It has great strength, particularly in his fore parts, in the muscles of its neck and jaws. It will carry off a sheep ,with ease, and with such rapidity that the shepherds have to use dogs to overtake it. The wolf is very cowardly by nature, and will not fight unless compelled to do so by the cravings of hunger, or to make good its retreat. Its sense of smell is so keen that it will scent a carcass at a distance of more than a league, and when leaving the woods it always travels against the wind. It prefers the animals destroyed by itself to those it may find dead ; but, in the absence of a live victim, will not disdain to partake of a defunct one. The color of the wolf differs according to the climate in which it is bred.

THE FALLOW DEER.

No two quadrupeds are more closely allied than the stag and fallow deer—the only apparent difference seems to be in their size and the form of their horns. The fallow deer is much smaller than the stag, and its horns, instead of being round, like those of the latter animal, are broad, palmated at the ends, and more profusely garnished with antlers. It sheds its horns annually, but they fall off later than those of the stag, and are renewed nearly at the same time. Notwithstanding the great similarity in their appearance and conformation, there are no two animals that keep more distinct, or avoid each other with more fixed animosity. They are never seen to herd in the same place; in fact it is a rare thing to find the fallow deer in a country where stags are numerous. It is found in almost every country in Europe, particularly in England, where it abounds; but it has never been seen in Russia, and rarely in the forests of Sweden. The fallow deer is easily tamed, feeds upon a variety of things which the stag refuses, and preserves its condition nearly the same through the whole year, although its flesh is considered much finer at particular seasons. It arrives at perfection in its third year, and lives to be about twenty.

THE PECCARY.

THIS animal is numerous in South America, and generally inhabits the forests, where it is met with in pairs. There are two species of the Peccary, the *collared* and the *white lipped*. The color of the former is usually of a yellowish grey, with the exception of the legs, which are nearly black, and it has an erect mane on the back and neck, composed of long and black bristles. The white-lipped is much larger than the collared, as it frequently reaches a length of three feet and a half, and a weight of one hundred pounds; while the collared rarely measures three feet or weighs more than fifty pounds.

In its native country, the Peccary prefers the mountainous parts to the low and level plains; it delights neither in marshes nor mud, as does our hog, but prefers roaming in the woods, where it subsists upon wild fruits, roots and vegetables. It is a

persistent enemy to the lizard, the toad, and all the serpent species, with which the country it inhabits abounds. As soon as it
perceives a serpent or a viper, it seizes it with its fore hoofs and
teeth, skins it in an instant and devours the flesh. The young
peccaries follow the dam and do not separate from her until they
have reached maturity. If taken when young they are easily
tamed and very soon lose their natural ferocity ; but they never
display any remarkable signs of docility, further than to be allowed to run about without any apprehension of dangerous results.

THE ANT - EATER.

THIS creature is a native of Brazil and Guiana, and is one of
three species, which are distinguished respectively as the
Great, the Middle, and the Lesser Ant-eater. The engraving represents the first named. This animal is nearly four feet
long, exclusive of the tail, which is two and a half, and covered
with rough hair exceeding twelve inches in length ; its head is
fourteen inches long, and its snout is of cylindrical shape, serving
as a sheath to its long and slender tongue, which always lies
folded in its mouth. Its legs are strong and only a foot high, the
fore ones being a trifle higher and more slender than those behind.
It swims over rivers, and subsists entirely on ants, which it col-
lects by thrusting its wire-like tongue into the ant-nests, and
having penetrated every part thereof withdraws it into its mouth
loaded with the insects.

The *Middle Ant-eater* measures one foot seven inches from nose to root of the tail, which is ten inches long, and with which it secures its hold in climbing trees by twisting it around the branches. It inhabits the same countries and procures its food in the same manner as the first named animal.

The *Lesser Ant-eater* is much smaller than the middle species, being not over six or seven inches in length, from nose to tail; its head is two inches long, and has a sharp pointed nose, inclining downwards; its fur is long, soft and silky, of a yellowish brown color. It inhabits Guiana, and climbs trees in quest of a species of ants that build their nests in the branches.

THE ICHNEUMON.

THIS is a small yellowish grey animal of the weasel kind, measuring from the tip of the nose to the extremity of the tail from thirty to forty inches, nearly half of which is occupied by the tail. Its eyes are of a bright red; its ears are small and rounded; its legs are short. This animal is domestic in Egypt, and, like the cat, is serviceable in destroying rats and mice; its instinct, however, is much stronger than that of the cat, for it hunts alike birds, quadrupeds, serpents, lizards and insects, and devours crocodiles eggs, which are deposited in the sand; nor does it spare chickens, if it can get at them. It lives by the side of rivers, inundations, and other waters, and, like the otter, it can swim and dive, and remain for a long while under the surface of the water. It quits its burrow to seek for prey at night, creeping cautiously like a serpent, and then erects itself on its hind legs to see whether danger or prey is near. Both the male and female ichneumon have a remarkable orifice, or opening, independent of the natural passages : it is a kind of pocket, into which an odoriferous liquor filters. Its nose is sharp and its mouth is narrow, which prevents its seizing or biting anything very large; but this defect is amply supplied by its courage, power, and agility. It easily strangles a cat, and will often fight with a dog, of whatever size it may be, and commonly gets the better of it. It can be easily domesticated, and proves to be more obedient and affectionate than the cat.

THE CIVET.

THIS animal belongs to the hottest countries in Asia and Africa; but it is capable of living in temperate, and even in cold countries, provided that it is carefully protected from the injuries of the air, and supplied with delicate and esculent food. The civet, or perfume, that is obtained from this animal is a substance of the consistence of honey or butter, of a clear yellowish or brown color, of a strong smell, and offensive when undiluted, but very agreeable when a small quantity is mixed with another substance. The manner in which the article is taken from this animal is peculiar. It is first placed into a long and narrow box, in which it cannot turn. The box is then opened from behind by the collector, who drags the animal by the tail, and keeps it in that position by a bar in front ; he then takes out the civet with a small spoon, carefully scraping the interior coats, under the tail, which secretes and contains it. The substance of perfume thus obtained is then put into a vessel, and every care is taken to keep it closely shut. The Civet is a wild, fierce animal, and, though occasionally tamed, is never familiar.

In appearance it resembles the polecat or the fox—is of an ash color tinged with yellow, marked with dusky spots disposed in rows; it is light and active and lives by prey, pursuing birds and the smaller quadrupeds which it is able to overcome. It is stealthy, like the fox, and often makes successful raids into yards and hen-roosts, for the purpose of carrying off poultry. Failing to procure animal food, it subsists upon roots and fruits. Its eyes shine at night, when it generally attacks its prey by surprise.

The quantity of perfume substance that one of these animals will yield depends greatly upon its appetite, and the quality of its nourishment. It produces more in proportion as it is more delicately and abundantly fed.

THE RACCOON.

THIS animal is a native of North America. It inhabits the southern parts of the fur districts, and is found as far north as the red river, in the Hudson Bay territory. In size it is about as large as a small badger, with a short and bulky body; its fur is fine, long and thick, blackish at the surface and grey towards the bottom; its head is like that of the fox, but its ears are round and shorter; its eyes are large, of a yellowish green, with a black and transverse stripe over them; its snout is sharp; has a thick tail, but tapering to a point, and is at least as long as the body; its fore legs are shorter than the hind ones, and both are supplied with five strong, sharp claws. In its wild state, the

Raccoon sleeps by day, and emerges from its retreat in the evening, prowling during the entire night in search of roots, fruit, birds, etc. The fur of the raccoon is used in the manufacture of hats, and its flesh, when it has been fed on vegetables, is said to make a savory dish. It can be easily tamed, when it is sportive and good natured; but is as mischievous as a monkey and never at rest. It is quite sensitive to ill treatment, and never forgets or forgives those from whom it is received.

THE TAPIR.

THE Tapir is a South American animal, and found principally in Brazil, Paraguay, and Guiana. It is about the size of a small cow. Like the hippopotamus, the tapir is amphibious, and has a thick skin, which is almost bullet proof. It is bulky in form, has a curved back, and a short thick neck covered with a bristly mane, which, near the head, is an inch and a half in length. Its head is moderately large, with roundish erect ears and small eyes. Its muzzle terminates in a kind of proboscis, about three inches long, which it can contract or extend at will: this member is used for the purpose of grasping and conveying food to its mouth. The tapir is nocturnal in its habits, seldom stirring about during the day time, and delights in the water, where it is oftener seen than upon land. Its haunts are in marshes, and seldom wanders far from the margins of rivers or lakes. When pursued or wounded by hunters, it plunges into the water, and remains under the surface until it reaches a sufficient distance to elude its pursuers. This animal, though chiefly living in the water, does not feed upon fish—he is a vegetarian, and subsists on fruits, shrubs, and grass. It is mild and timid in its nature, and flies from every threatened attack or danger; but when escape is impossible it will make a vigorous resistance. Its attitude is that of sitting on the posterior parts, like a dog; and its voice is an excellent imitation of a whistle. The tapir when tamed is gentle and docile, and its flesh is considered wholesome.

THE OTTER.

ALTHOUGH this animal is not considered by naturalists as altogether amphibious, it is, notwithstanding, capable of remaining for a long time under water, where it pursues its prey with great facility. The legs of the otter are very short but remarkably strong, broad and muscular; on each foot are five toes, connected by membranes like those of a water fowl; its head is broad, of an oval form, and flat on the upper part; the body is long and round, and the tail tapers to a point; the eyes are brilliant, and placed in such a manner that the animal can see every object that is above it, which gives it a singular appearance, very much resembling an eel or an asp; the ears are short and their orifice narrow. The fur of the otter, which is valuable, is of a dark brown color, with two small light spots on each side of

the nose, and another under the chin. The otter makes its bed
in a secluded spot by the side of a river or lake, under a bank,
where it can have easy access to the water, whither it flies upon
the least alarm. It destroys great quantities of fish, pursuing
which it often swims against the stream. As soon as it catches
a fish he takes it to the shore and devours it, and then returns to
the water in search of more. Otters are found in most parts of
the world, with but little variation in their conformation.

THE SLOTH.

THIS is the most sluggish and inactive of all animals, and,
judging from its appearance, it seems to be the most help-
less and wretched. Its every motion seems to be the effect
of painful exertions, which hunger alone is capable of exciting.
It takes up its abode chiefly in trees, where it will remain until it
strips every vestage of verdure therefrom—sparing neither fruit,
blossom, nor leaf; after which it is said to devour even the bark.
Being unable to descend, it throws itself to the ground, where it
remains at the bottom of the tree till the pangs of hunger again
spurs it to renew its toils in search of subsistence. Its motions
are accompanied with a piteous and lamentable cry, which ter-
rifies beasts of prey, and proves its best defence. Though slow,
awkward, and almost incapable in its motion, the Sloth is strong,
remarkably tenacious of life, and capable of enduring a long ab-
stinence from food. There are two kinds of sloths, and are distinguished by the number of their claws. The one, called the
Ai, is about the size of a fox, and has three long claws on each
foot; its legs are clumsy and awkwardly placed,—the fore legs
being longer than the hinder add to the difficulty of its progress-
ive motion; its body is covered with a rough coat of long hair, of

a lightish brown color, mixed with white, and has a black line down the middle of the back; its face is naked and of a dirty white color; its eyes are small, black and heavy. This species is found only in South America. The other, called the *Unau*, has but two claws on each foot; its head is short and round, like that of a monkey; its ears are short and is without a tail. It is found in the Island of Ceylon and South America.

MONKEYS.

THE tailless *Macauco* is a native of Bengal and the island of Ceylon. The head of this animal is small, with pointed nose; its eyes are edged with a circle of white, which is also surrounded with another of black; its body is covered with a short, silky fur, of a reddish ash color. Its length from the nose to the rump is sixteen inches. It is a very inactive animal and slow in it motions; is very tenacious of its hold, and makes a plaintive noise. It dwells in the woods, and subsists upon fruits, small birds, etc.

The *Mongooz* is found in Madagascar and the isles adjacent. It sleeps in trees, and feeds on fuits; is playful and good natured.

Its fur is fine, soft and woolly, and of a deep brownish ash color; its eyes are of a beautiful orange color, surrounded with black; its ears are short, cheeks white, and nose black; tail very long, and covered with hair of the same color as the body; its hands and feet are naked, and of a dusky color.

THE MANDRILL.

THIS Baboon is found on the Gold Coast and other southern parts of Africa, where it is called by the negroes *boggo*, and by the Europeans, *mandrill;* it also bears the name of the ribbed-nose baboon. It is an ugly, disgusting looking animal, and remarkable for its variety of color, its singularity of appearance, its great strength, and its unconquerable savageness. When standing upright, the Mandrill is from three to four feet high. It has a projecting forehead, under which are two small and vivid eyes, placed so near to each other that their position alone gives its physiognomy an appearance of ferocity. An enormous muzzle, indicative of brutal passions, terminates in a broad and rounded extremity, of a fiery red color, from which continually oozes a mucous humor. The cheeks, swollen and burrowed, are naked and of a deep blue color. Round the neck the hair is long, on the sides of the head it joins at the top, and the whole terminates in a somewhat pointed form. Each hair of the body is annulated with black and yellow, so that the fur has a greenish brown hue. Its voice has a resemblance to the lion's roar. - No art or kindness can subdue the brutal propensities of this animal, and its immense strength renders it an object of perpetual dread to its keepers. It usually feeds on fruit and nuts, but, though not, strictly speaking, a carnivorous animal, will eat cooked meat.

THE DOG - FACED BABOON.

*T*HIS animal inhabits the hottest parts of Africa and Asia, where it is found in great number. It is about five feet high, exceedingly strong, vicious and impudent, and is distinguished by a longer tail than the rest of its kind. Its head is very large, with a long and thick muzzle; eyes small, face naked and of an olive color; the hair on its forehead is separated in the middle and hangs down on each side of the face, and from thence down the back to the waist; the hair on the lower part of the body is short, and its buttocks are bare and red.

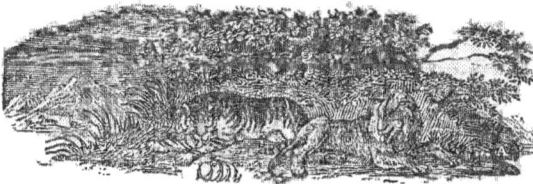

www.ingramcontent.com/pod-product-compliance
Lightning Source LLC
Chambersburg PA
CBHW022027190326
41519CB00010B/1624